le 5

Discovery Education | SCIENCE TECHBOOK

California
Unit 3
Interacting Earth Systems

Discovery EDUCATION

To obtain permission(s) or for inquiries, submit a request to:

Discovery Education, Inc.
4350 Congress Street, Suite 700
Charlotte, NC 28209
800-323-9084
Education_Info@DiscoveryEd.com

ISBN 13: 978-1-68220-551-8

Printed in the United States of America.

1 2 3 4 5 6 7 8 9 10 BRP 28 27 26 25 A

Acknowledgments

Acknowledgment is given to photographers, artists, and agents for permission to feature their copyrighted material.

Cover and inside cover art: ver0nicka / Shutterstock.com

Table of Contents

Unit 3: Interacting Earth Systems

Concept 3.3 Protecting Earth's Resources

Unit Wrap-Up

Grade 5 Resources

Dear Parent/Guardian,

This year, your student will be using Science Techbook™, a comprehensive science program developed by the educators and designers at Discovery Education and written to the California Next Generation Science Standards (NGSS). The California NGSS expect students to act and think like scientists and engineers, to ask questions about the world around them, and to solve real-world problems through the application of critical thinking across the domains of science (Life Science, Earth and Space Science, Physical Science).

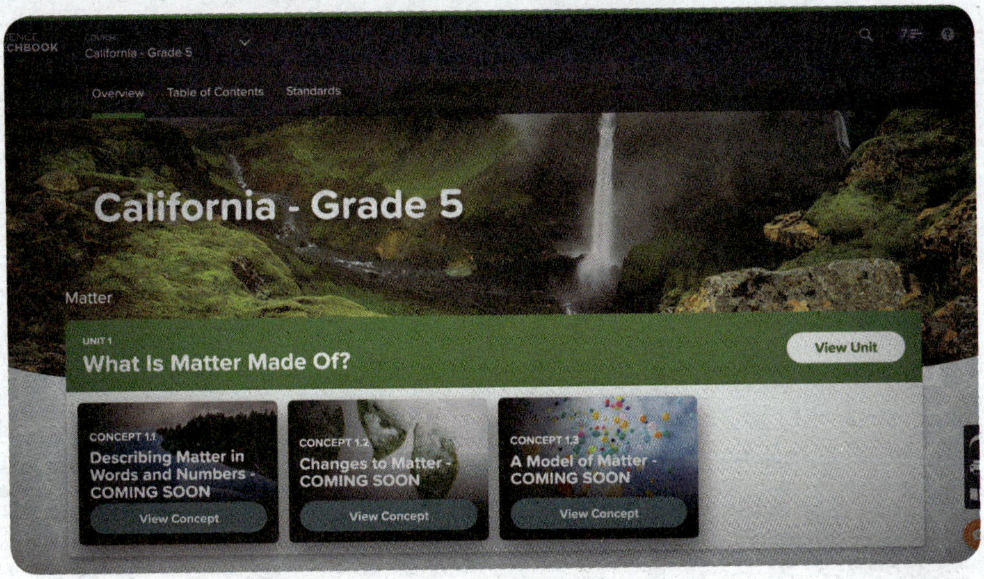

Science Techbook is an innovative program that helps your student master key scientific concepts. Students engage with interactive science materials to analyze and interpret data, think critically, solve problems, and make connections across science disciplines. Science Techbook includes dynamic content, videos, digital tools, Hands-On Activities and labs, and game-like activities that inspire and motivate scientific learning and curiosity.

You and your child can access the resource by signing in to www.discoveryeducation.com. You can view your child's progress in the course by selecting the Assignment button.

Science Techbook is divided into units, and each unit is divided into concepts. Each concept has three sections: Wonder, Learn, and Share.

Units and Concepts Students begin to consider the connections across fields of science to understand, analyze, and describe real-world phenomena.

Wonder Students activate their prior knowledge of a concept's essential ideas and begin making connections to a real-world phenomenon and the **Can You Explain?** question.

Learn Students dive deeper into how real-world science phenomenon works through critical reading of the Core Interactive Text. Students also build their learning through Hands-On Activities and interactives focused on the learning goals.

Share Students share their learning with their teacher and classmates using evidence they have gathered and analyzed during Learn. Students connect their learning with STEM careers and problem-solving skills.

Within this Student Edition, you'll find QR codes and quick codes that take you and your student to a corresponding section of Science Techbook online. To use the QR codes, you'll need to download a free QR reader. Readers are available for phones, tablets, laptops, desktops, and other devices. Most use the device's camera, but there are some that scan documents that are on your screen.

For resources in California Science Techbook, you'll need to sign in with your student's username and password the first time you access a QR code. After that, you won't need to sign in again, unless you log out or remain inactive for too long.

We encourage you to support your student in using the print and online interactive materials in Science Techbook, on any device. Together, may you and your student enjoy a fantastic year of science!

Sincerely,

The Discovery Education Science Team

Unit 3
Interacting Earth Systems

Water Solutions: Recycling Wastewater

Earth's freshwater supplies are shrinking. One solution to this problem is to recycle wastewater. California is encouraging homeowners and commercial buildings to recycle their wastewater. Throughout this unit, you will explore a variety of phenomena that arise from the interaction of the biosphere and hydrosphere. By the end of the unit, you will be able to use conceptual models that you build to understand how humans impact and protect these particular Earth systems.

Quick Code:
ca5505s

Water Solutions: Recycling Wastewater

Think About It

Look at the photograph. **Think** about the following questions.

- How can we represent systems as complicated as the entire planet?
- Where does my tap water come from, and where does it go?
- How much water do we need to live and to irrigate plants?
- How much water do we have?
- What can we do to protect Earth's resources?

Water Solutions

Solve Problems Like a Scientist

Quick Code: ca5506s

Case Study: Tracking Water Pollution

In this project, you will apply your knowledge of water ecosystems and how Earth systems interact to analyze data about the health of watersheds and develop a model of a water filtration device.

Tracking Water Pollution

SEP Developing and Using Models

SEP Analyzing and Interpreting Data

SEP Using Mathematics and Computational Thinking

CCC Systems and System Models

Ask Questions About the Problem

You are going to design and test a model of a water filtration system using what you know about Earth systems and resources. **Write** some questions you can ask to learn more about the problem. As you learn about how Earth's systems interact and how humans can protect resources, **write** down answers to your questions.

CONCEPT
3.1

Hydrosphere and Biosphere Interactions

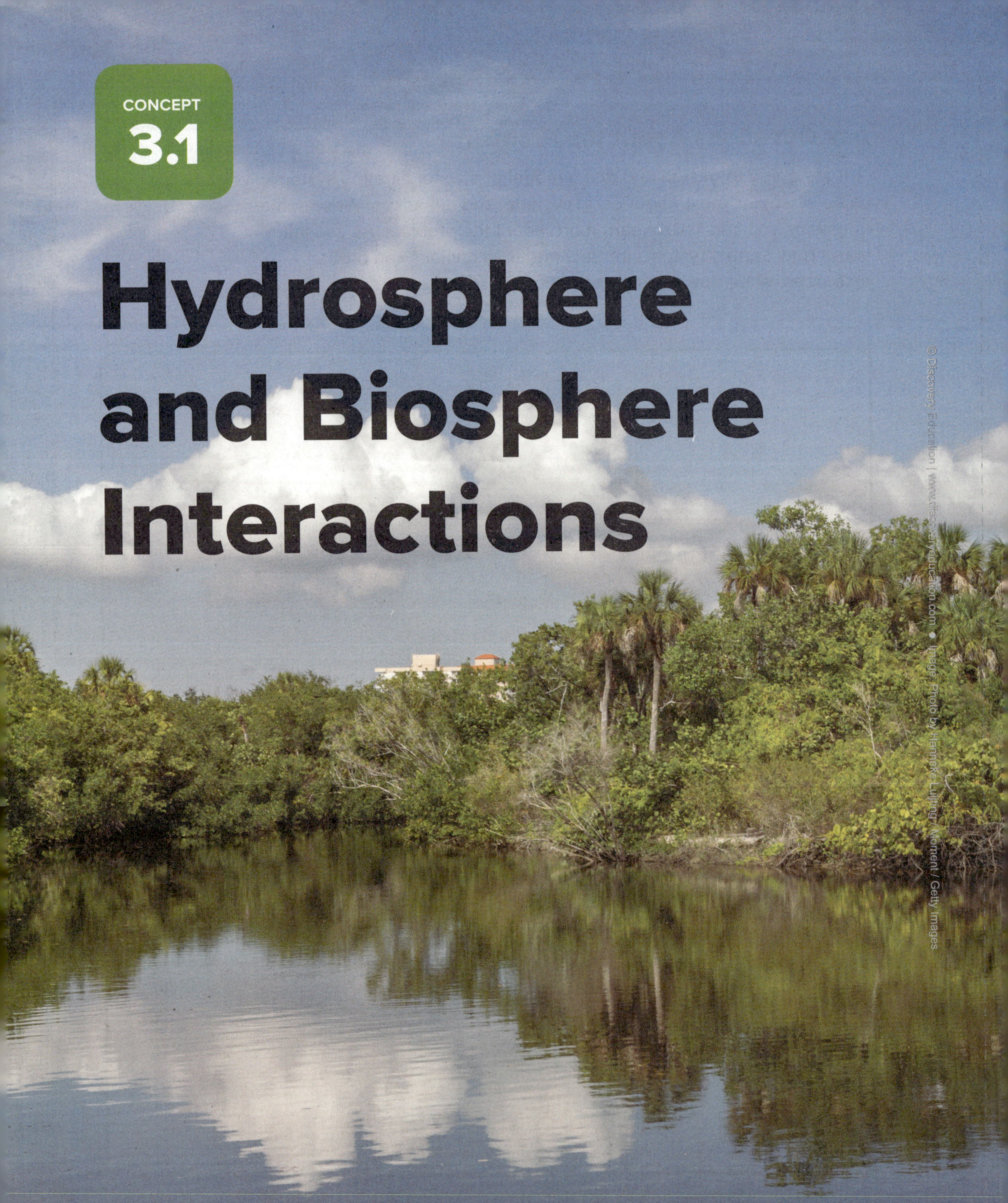

Student Objectives

By the end of this lesson:

- [ ] I can develop a model that shows how the geosphere, biosphere, hydrosphere, and atmosphere interact.
- [ ] I can explain how plants and animals have internal and external structures and behaviors that support survival, growth, behavior, and reproduction.

Key Vocabulary

- [ ] abiotic
- [ ] abyssal zone
- [ ] aquatic
- [ ] Arctic
- [ ] atmosphere
- [ ] biosphere
- [ ] biotic
- [ ] Earth
- [ ] ecosystem
- [ ] estuary
- [ ] geosphere
- [ ] hydrosphere
- [ ] natural resource
- [ ] ocean
- [ ] organism
- [ ] phenomenon
- [ ] rely
- [ ] substance
- [ ] terrarium
- [ ] watershed

Quick Code:
ca5508s

Activity 1

Can You Explain?

How does Earth's biosphere interact with Earth's hydrosphere?

Quick Code:
ca5509s

Activity 2

Ask Questions Like a Scientist

Quick Code:
ca5510s

Water for Plants

Look at the image. **List** the reasons you think water is important. You can arrange your list into categories.

Let's Investigate Water for Plants

Activity 3

Observe Like a Scientist

Quick Code: ca5512s

The Importance of Water for Life on Earth

Watch the three videos. As you watch, **complete** the statements.

Introduction: Plants

Our Blue Planet

Water: Earth's Important Resource

I wonder ______________________________.

I wonder ______________________________.

I wonder ______________________________.

I wonder ______________________________.

I wonder ______________________________.

Activity 4

Evaluate Like a Scientist

Quick Code: ca5514s

What Do You Already Know About Hydrosphere and Biosphere Interactions?

Forms of Water

Write each word from the word bank next to the statement that best describes it.

estuary	groundwater	lake
ocean	river	runoff

A body of water surrounded by land that is often fresh but sometimes salty. ____________________

Water that flows from an area of high altitude to an area with lower altitude in a definite channel. ____________________

Water from rain or snow that moves into rivers or the ground.

____

A very large body of saltwater. ____________________

The mouth of a river that meets the ocean. ____________________

Water that fills billions of tiny spaces underground.

____

Plant Use

Read the statement and **answer** the questions.

Your friend says that the main use for plants is for decorative gardens around homes and in parks. Is your friend correct? Why or why not?

Renewable Resource

Think about what you know about plants and resources. Then, **answer** the question.

Why are plants considered a renewable resource, and how are they renewable?

How Do Earth's Biosphere and Hydrosphere Interact with Each Other?

Activity 5

Observe Like a Scientist

Quick Code: ca5515s

Food Chains and Food Webs

Look at the image. **Watch** the video. As you watch, **observe** and **note** relationships and dependencies among organisms. Then, **complete** the activity.

Coral Reef Food Web

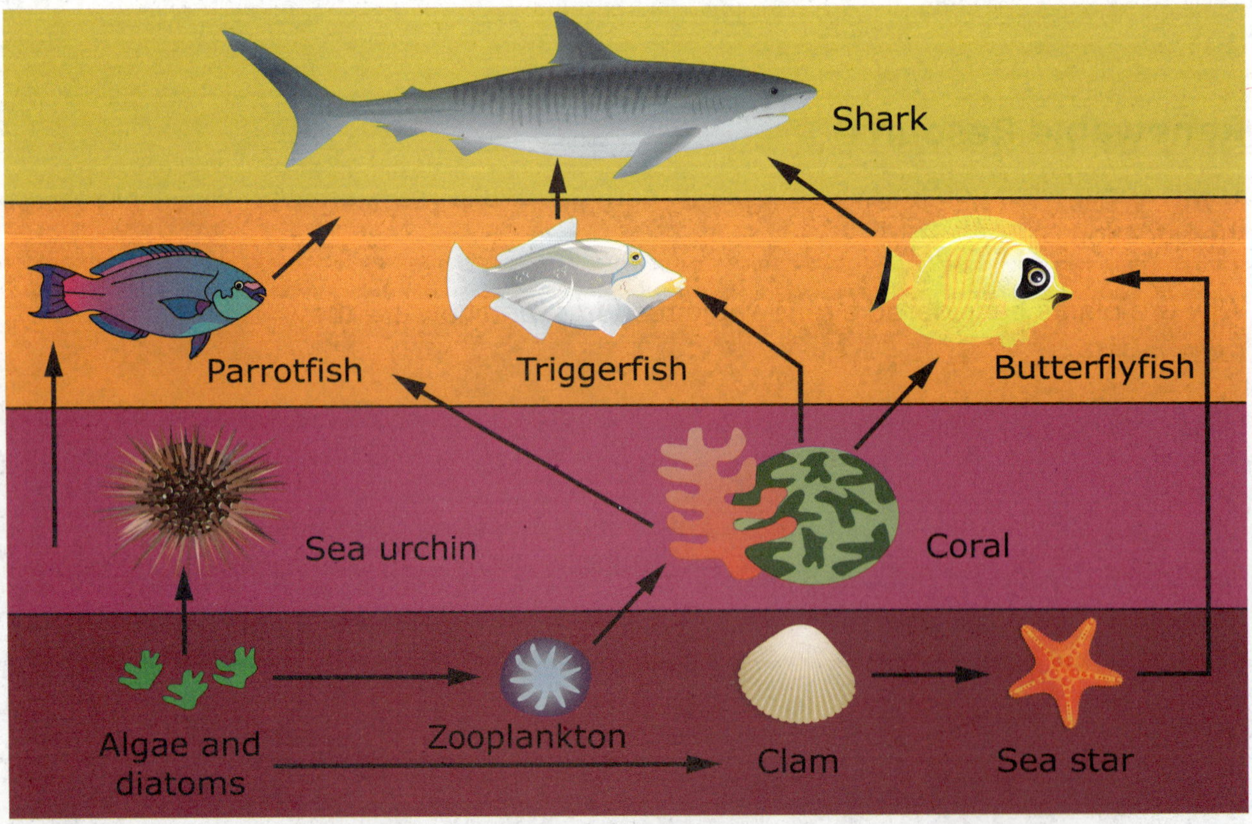

Food Chain

Now, **draw** a food chain or food web based on familiar plants and animals, such as those that might be in a classroom **terrarium** or found in a fictional text.

Activity 6

Think Like a Scientist

Quick Code: ca5517s

What's in Our Schoolyard?

In this investigation, you will explore your schoolyard. You will then list and categorize the living and nonliving objects you find there.

What Will You Do?

Explore the schoolyard. **List** the living and nonliving objects you find there. Then, **categorize** your list using the graphic organizer.

Geosphere (Land)	Hydrosphere (Water)	Atmosphere (Air)	Biosphere (Life)

SEP Planning and Carrying Out Investigations

CCC Patterns

Think About the Activity

What patterns do you see in your observations?

In what ways are the objects in each type of system necessary to sustain life?

Activity 7

Analyze Like a Scientist

Quick Code: ca5518s

Earth's Systems

Read the text. Then, **choose** one of the systems and **draw** a picture including living and/or nonliving objects that make up the system.

Earth's Systems

Over time and after observing many living and nonliving things on **Earth**, scientists have settled on four major categories to sort objects and **phenomena** into common groups or systems. One system is the **geosphere**. This system contains things such as rocks, minerals, landforms, soil, and even molten rocks inside Earth. Another system is the **hydrosphere**. This contains all water on Earth, such as the **oceans**, rivers, groundwater, and glaciers. The third system is the **atmosphere**. The atmosphere contains all the gases that surround Earth. We typically call these gases air. Air is a mixture of many kinds of gases. The last system is the **biosphere**. This contains all living **organisms**, including humans. Together, these four systems make up the Earth system.

Scientists have found it useful to look at and study how these four different systems work together and interact. When the hydrosphere interacts with the geosphere, you have phenomena such as erosion and lake formation. When the atmosphere interacts with the biosphere, you see plants giving off water and gases as waste products. When the geosphere interacts with the biosphere, you find soil providing nutrients for plants. In each of these phenomena, there is an exchange of energy and matter.

Activity 8

Observe Like a Scientist

Quick Code: ca5520s

Characteristics of the Hydrosphere and Biosphere

Watch the two videos. As you watch, **list** the characteristics of each system in the graphic organizer. Then, **write** as many statements as you can about how the hydrosphere and biosphere interact.

What Is the Biosphere?

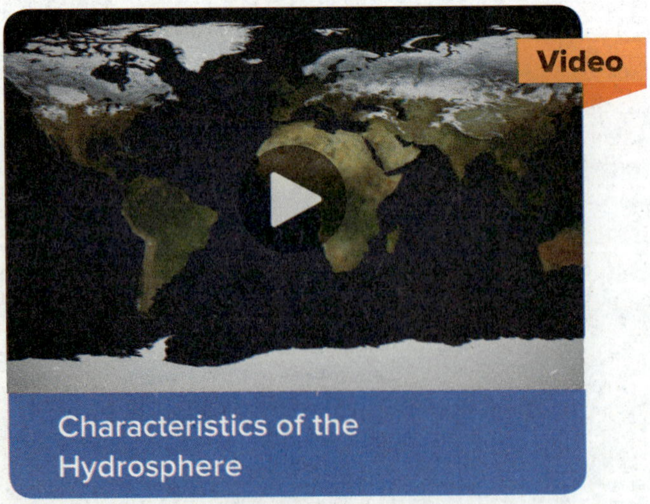

Characteristics of the Hydrosphere

Biosphere and Hydrosphere

Biosphere	Hydrosphere

Activity 9

Observe Like a Scientist

Quick Code: ca5521s

Identify Interactions

Look at the images. For each image, **write** the interactions you see occurring in it. Then, **discuss** with a partner the interactions you see.

SEP Constructing Explanations and Designing Solutions

Activity 10

Evaluate Like a Scientist

Quick Code: ca5522s

Hydrosphere or Biosphere?

Write the observations in the correct column.

- saltwater in the ocean
- a spring flows out from under a rock
- ants feast on a sandwich
- water evaporates from a pond
- bees pollinate a flower
- a hawk spots its prey
- an iceberg breaks off from its glacier
- a frog lays eggs

Hydrosphere	Biosphere

Activity 11

Observe Like a Scientist

Quick Code: ca5523s

Coral Reefs

Look at the image. **Write** a short story about the interactions shown in the image.

Coral Reefs

SEP Obtaining, Evaluating, and Communicating Information

How Do Water Ecosystems Interact between the Biosphere and Hydrosphere?

Activity 12

Analyze Like a Scientist

Quick Code: ca5524s

Types of Aquatic Ecosystems

Read the text. As you read, **highlight** the components in each system. Then, **discuss** with your class what would happen if these components increased or decreased.

Types of Aquatic Ecosystems

Water **ecosystems** are also called **aquatic** ecosystems. There are many different types of aquatic ecosystems. They can be classified in different ways.

Some aquatic ecosystems are in saltwater. The salty oceans cover a large portion of Earth's surface. Ocean ecosystems include shallow areas, such as coral reefs and intertidal zones. The intertidal zone is the area along the ocean between the high tide and the low tide. Ocean ecosystems also include its deepest areas, called the **abyssal zone**. The abyssal zone is so deep that sunlight cannot reach it. The ocean contains a great diversity of organisms.

CCC **Cause and Effect**

Types of Aquatic Ecosystems *cont'd*

Ecosystems found along the edges of the oceans, such as estuaries, have a mixture of saltwater and freshwater. Some lakes, such as the Great Salt Lake in Utah, also have saltwater ecosystems. This lake is far from the ocean, but it has a high concentration of natural salts. Consequently, it is too saline to support fish and most other aquatic species. Several types of algae live in the lake. Brine shrimp and brine flies can tolerate the high salt content, and they feed on the algae.

Water Ecosystem

Antarctic Aquatic System

Other aquatic ecosystems are freshwater. Freshwater ecosystems include ponds and lakes. In many ponds and lakes, the water is present all year. Other ponds and lakes dry up in the hot, summer months. The plants and animals that live there must adapt to the change. Water ecosystems are also found in flowing bodies of water, such as streams and rivers. Different plants and animals thrive in water that is always moving. Streams and rivers also connect other bodies of water, such as lakes and oceans.

Activity 13

Observe Like a Scientist

Water Ecosystems

Quick Code: ca5526s

Complete the interactive. **Build** your ecosystem. Then, **record** your data in the graphic organizer.

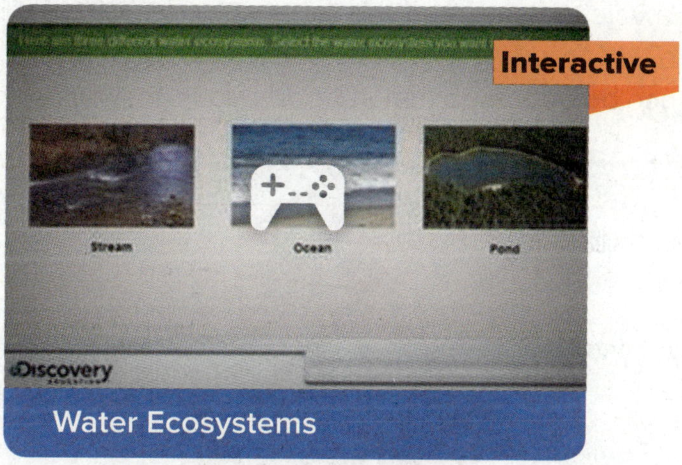

Water Ecosystems

Ecosystem	Type of Water	Water Movement	Species
Pond			
Stream			
Ocean			

Activity 14

Evaluate Like a Scientist

Quick Code: ca5525s

What Is a Volcano? and Affecting Earth's Spheres

What Is a Volcano?

Look at the image of a volcano and **read** the list of spheres. **Circle** the sphere that includes a volcano. **Think** about what you read in Earth's Systems. What evidence do you have that supports your answer?

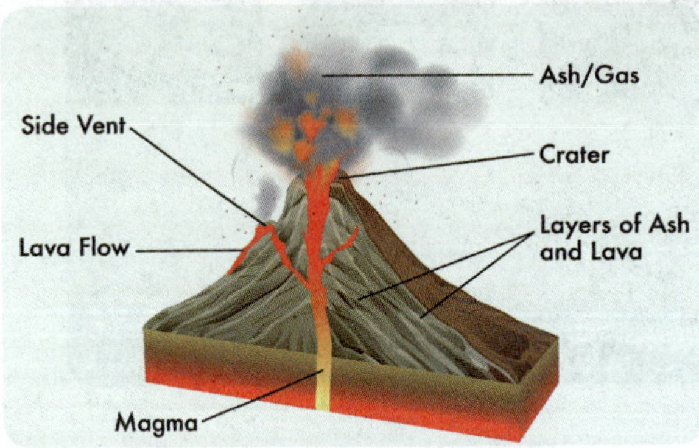

A. atmosphere **B.** biosphere **C.** geosphere **D.** hydrosphere

Affecting Earth's Spheres

Watch the video Notable Volcanic Eruptions. A volcanic eruption can impact all of Earth's spheres.

Notable Volcanic Eruptions

Read the list of effects caused by a volcanic eruption. **Match** each effect with the Earth spheres that would be impacted. Each effect will have connections to more than one Earth sphere.

Effects	Earth spheres
steam produced by lava entering ocean	atmosphere
snow being melted due to heat from an eruption	hydrosphere
trees being burned by hot lava	biosphere
homes damaged by flowing lava	geosphere
ash and gases released during an eruption	

Activity 15

Record Evidence Like a Scientist

Quick Code: ca5528s

Water for Plants

Now that you have learned about water for plants, look again at the image of water for plants. You first saw this in Wonder.

Let's Investigate Water for Plants

SEP Constructing Explanations and Designing Solutions

Talk Together

How can you describe water for plants now?

How is your explanation different from before?

Look at the Can You Explain? question. You first read this question at the beginning of the lesson.

Can You Explain?

How does Earth's biosphere interact with Earth's hydrosphere?

Now, you will use your new ideas about the biosphere and hydrosphere to answer a question.

1. **Choose** a question. You can use the Can You Explain? question or one of your own. You can also use one of the questions that you wrote at the beginning of the lesson.

My Question

2. Then, **use** the checklist to help you **answer** the question.

To **plan** your scientific explanation, **look** at the checklist.

- ☐ Did you include an example of how living or nonliving objects relate to one another?
- ☐ Does your evidence include data? If you included data, did you:
- ☐ Show the measurements?
  - ☐ Describe how you collected the data?
  - ☐ Does each piece of your evidence support your claim?

Now, **write** your scientific explanation.

Earth's hydrosphere and biosphere interact by...

STEM in Action

Quick Code: ca5529s

Activity 16
Analyze Like a Scientist

North America's Treasure

Read the text and **look** at the image. Then, **complete** the activity.

North America's Treasure

The Chesapeake Bay is a large **estuary**. It runs from Maryland to Virginia. The Chesapeake Bay is the largest estuary in North America! Fifty major rivers and streams and many smaller water ecosystems flow into the bay. The **watershed** area of the bay is an area of land where all of the water that falls onto and drains from the land flows to the bay. An estimated 17 million people regularly spend time in the Chesapeake Bay watershed.

The Chesapeake Bay

SEP Obtaining, Evaluating, and Communicating Information

North America's Treasure *cont'd*

As you can imagine, the Chesapeake Bay is home to a great variety of organisms. Protecting this ecosystem from harmful practices like overfishing and pollution is important. The Chesapeake Bay Foundation (CBF) is an organization that works to protect the bay. One way they do this is through their education programs. Many CBF educators are scientists with degrees in marine biology. They have to know a lot about the water ecosystems that are part of the watershed. CBF educators teach students about the bay ecosystem. They teach about the importance of protecting it. Many of the educators teach in the field. For example, in one education program, educators and students spend the day on a sailboat learning about oysters, which are shelled animals that live in the Chesapeake Bay ecosystem.

Phosphorus pollution is a form of pollution in which too much of the nutrient phosphorus enters a body of water. It causes a harmful imbalance within the water ecosystem. Two sources of phosphorus are plant fertilizers used in agriculture (farming) and wastewater. It can also enter the air through water vapor as part of the water cycle. The graphs show data about the ways in which phosphorus pollution enters the bay in two areas and in the entire watershed.

Use the data to **make** three claims about phosphorus pollution in the Chesapeake Bay.

Phosphorus Pollution to the Chesapeake Bay, by Source

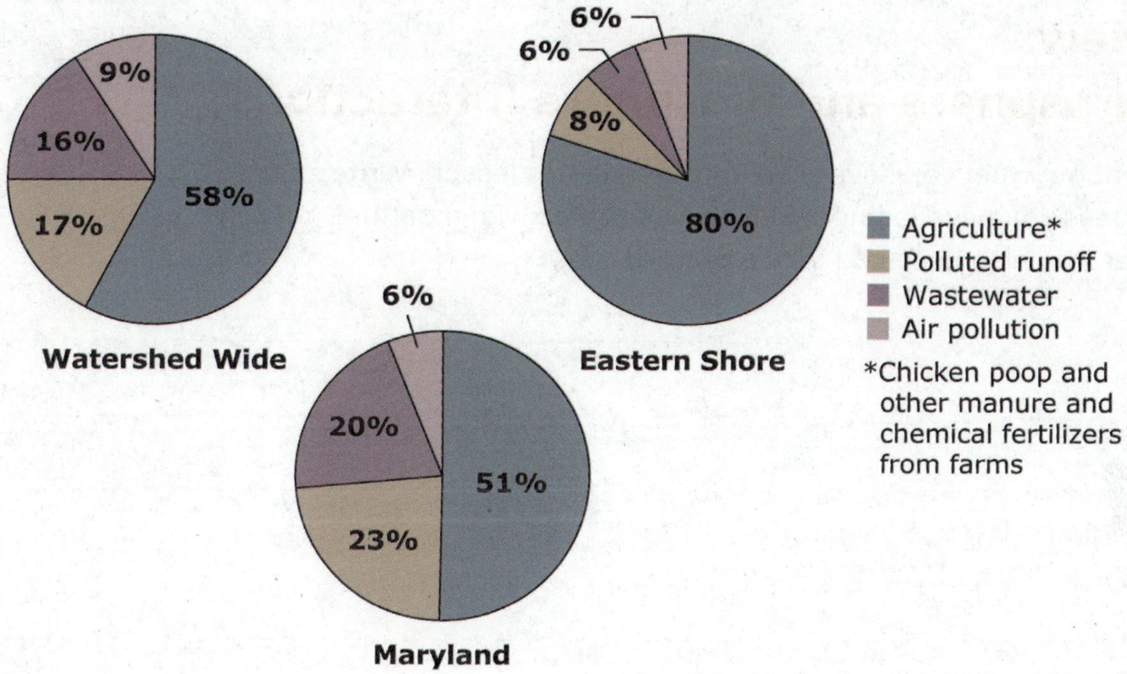

Source: Based upon Chesapeake Bay Program Chesapeake Bay Watershed 5.3.2

Activity 17

Evaluate Like a Scientist

Quick Code: ca5533s

Review: Hydrosphere and Biosphere Interactions

Think about what you have read and seen in this lesson. **Write** some core ideas you have learned. Review your notes with a partner. Your teacher may also have you take a practice test.

SEP Obtaining, Evaluating, and Communicating Information

Talk Together

Think about what you saw in Get Started. Use your new ideas to discuss the importance of water for plants and other systems.

CONCEPT
3.2

Water as a Valuable Natural Resource

Student Objectives

By the end of this lesson:

- [ ] I can graph relative proportions of water found in different forms on Earth.
- [ ] I can develop a model that describes the patterns in water distribution on Earth.
- [ ] I can develop a model of the hydrosphere that describes how its components interact.
- [ ] I can argue from evidence for the most important source of freshwater.
- [ ] I can compare multiple ways to conserve freshwater.

Key Vocabulary

- [ ] crater
- [ ] freshwater
- [ ] glacier
- [ ] mineral
- [ ] precipitation
- [ ] refuse
- [ ] resource
- [ ] river
- [ ] salt
- [ ] saltwater
- [ ] stream
- [ ] volcano
- [ ] water

Quick Code:
ca5535s

Activity 1

Can You Explain?

Why is water such an important natural resource?

Quick Code:
ca5537s

Activity 2

Ask Questions Like a Scientist

Quick Code:
ca5538s

Our Blue Planet

Watch the video. Then, **complete** the activity that follows.

Video

Let's Investigate Our Blue Planet

SEP **Asking Questions and Defining Problems**

Write about places you have been where water was present.

Activity 3

Observe Like a Scientist

Quick Code: ca5539s

The Importance of Water

Watch the video. Then, **draw** or **describe** ways that humans, animals, and plants use water.

The Importance of Water

Humans

Animals

Plants

Activity 4

Observe Like a Scientist

Quick Code: ca5540s

Wonder About Water

Look at the image. Then, **complete** the graphic organizer.

Earth from Space

I see...

I think...

I wonder...

Activity 5

Evaluate Like a Scientist

Quick Code: ca5541s

What Do You Already Know About Water as a Valuable Natural Resource?

Conserving Water

Underline the **three** most practical ways to conserve water.

Reduce the number of times you wash a car.

Drink more juice instead of water.

Turn off the faucet while you brush your teeth.

Only go swimming in saltwater.

Limit the amount of steamed vegetables you eat.

Only run the dishwasher when it is completely full.

Groundwater

The diagram shows an area that will be developed into a neighborhood. Several houses will be built. The developers need to add a well, which collects and pumps freshwater for people to use. **Circle** the part of the diagram where the well should collect freshwater.

Activity 6

Analyze Like a Scientist

How Is Water Distributed?

Quick Code: ca5542s

Read the text. As you read, **highlight** information you can use as evidence to support your initial ideas on how to answer the Can You Explain? question or your own question you generated earlier.

How Is Water Distributed?

Did you know that your body is made up of about 65 percent **water**? Like all organisms, you need water to survive. Water is part of what makes life as we know it possible on Earth. Think about how often you drink water, wash with water, and use water throughout the day. Odds are you get a great deal of the water you use from the faucets in your house. Much of that water comes from **freshwater** sources. These sources include **rivers**, lakes, **streams**, aquifers, and even human-made reservoirs. However, most of Earth's water is not freshwater.

Water covers more than 70 percent of Earth's surface. More than 97 percent of that surface water is found in Earth's oceans. Geographers have divided the massive body of water that surrounds all the continents into five distinct oceans. These are the Arctic Ocean, the Atlantic Ocean, the Indian Ocean, the Pacific Ocean, and the Southern Ocean. Oceans are **saltwater** bodies, as are many inland seas, such as the Mediterranean Sea. Saltwater contains **salt** and other **minerals** that make the water unsuitable for drinking. Ocean water must undergo an extensive and costly process to remove the salt from it for people to be able to drink it.

Activity 7

Evaluate Like a Scientist

Quick Code: ca5543s

Distribution of Earth's Waters

Most of Earth's surface is covered by water. Most of that water is in the oceans and salty, and only a small amount of that water is freshwater. Here are the relative amounts of each type of water.

Ocean Water	Freshwater
97%	3%

Using the table provided, **create** a pie graph to represent this information.

SEP Using Mathematics and Computational Thinking

Activity 8

Observe Like a Scientist

Quick Code: ca5544s

Waters of Earth

Complete the interactive. **Record** your observations in the data table that follows.

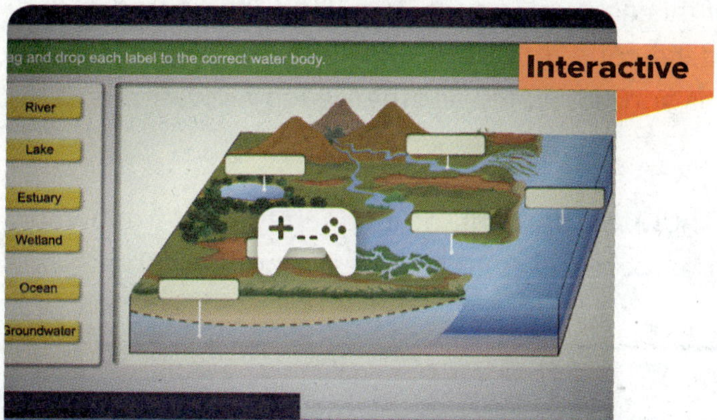

Waters of Earth

Water Body	Type of Water	Where It Forms	Other Information

Activity 9

Evaluate Like a Scientist

Quick Code: ca5545s

Earth's Freshwater

Read the information about Earth's freshwater.

Scientists estimate that Earth has about 35 million cubic kilometers (km^3) of freshwater. Of this amount, about 24 million km^3 is trapped in the ice of **glaciers** and ice caps. About 10.5 million km^3 is groundwater. About 0.5 million km^3 is surface water or other (such as water in Earth's atmosphere).

Create a pie chart to show the percentage of each type of freshwater on Earth. **Round** each percentage to the nearest whole number.

SEP Using Mathematics and Computational Thinking

Activity 10

Think Like a Scientist

Quick Code: ca5546s

Local Freshwater: Part 1

In this investigation, you will **conduct** safe searches on the Internet to **answer** the question that follows.

What Will You Do?

Where does your local tap water come from? **Use** the following terms: [city/town + municipal water supply] OR [county + tap water supply]

SEP **Obtaining, Evaluating, and Communicating Information**

Activity 11

Think Like a Scientist

Quick Code: ca5547s

Local Freshwater: Part 2

In this investigation, you will **conduct** safe searches on the Internet to **answer** the question that follows.

What happens to our wastewater? **Use** the following terms: [city/town + wastewater] OR [county + tap wastewater]

SEP Obtaining, Evaluating, and Communicating Information

Activity 12

Think Like a Scientist

Quick Code: ca5548s

Local Freshwater: Part 3

In this investigation, you will **conduct** safe searches on the Internet to **answer** the question that follows.

Which human activities are the primary users of local freshwater sources? **Use** the following terms: [human actions + water use]

Activity 13

Think Like a Scientist

Quick Code: ca5549s

Local Freshwater: Part 4

Review the information you gathered about your local freshwater.
Use the information to **answer** the questions.

Think About the Activity

What are the most important sources of freshwater, generally? What are the most important sources in your area?

How can we reduce our freshwater usage individually and as a society? Why might this be desirable?

SEP **Obtaining, Evaluating, and Communicating Information**

Activity 14

Evaluate Like a Scientist

Quick Code: ca5550s

Sources of Freshwater

Use what you have learned about sources of freshwater to **answer** the question.

So far, you have learned about several sources of freshwater, including lakes, rivers, streams, and ponds. What is the most important source of freshwater? Support your argument using at least two reasons.

SEP **Engaging in Argument from Evidence**

Why Is It Important to Conserve Freshwater?

Activity 15

Observe Like a Scientist

Freshwater: A Precious Resource

Quick Code: ca5551s

Watch the video. As you watch, **complete** the information on the four "sticky notes."

Freshwater: A Precious Resource

SEP Engaging in Argument from Evidence

1 factual statement

1 question

1 vocabulary word
(with definition at the bottom)

1 sketch/drawing that explains content from the video

Activity 16

Record Evidence Like a Scientist

Quick Code: ca5552s

Our Blue Planet

Now that you have learned about water as a natural resource, watch the video of our blue planet again. You first saw this in Wonder.

Let's Investigate Our Blue Planet

Talk Together

How can you describe our blue planet now?

How is your explanation different from before?

SEP **Constructing Explanations and Designing Solutions**

Look at the Can You Explain? question. You first read this question at the beginning of the lesson.

Can You Explain?

Why is water such an important natural resource?

Now, you will use your new ideas about our blue planet to answer a question.

1. **Choose** a question. You can use the Can You Explain? question or one of your own. You can also use one of the questions that you wrote at the beginning of the lesson.

My Question

2. For help answering the above question, **use** the checklist that follows. Then, **use** the checklist to plan your scientific explanation.

To plan your scientific explanation, look at the checklist.

- ☐ Did you include examples of the uses of water?
- ☐ Does your evidence include data? If you included data, did you:
  - ☐ Show the measurements?
  - ☐ Describe how you collected the data?
- ☐ Does each piece of your evidence support your claim?

Now, **write** your scientific explanation.

Water is an important natural resource because...

STEM in Action

Quick Code: ca5555s

Activity 17

Analyze Like a Scientist

There's a Bug in My Water!

Read the text and **watch** the video. Then, **discuss** the questions.

There's a Bug in My Water!

As you've learned, freshwater is an important **resource** to all life on Earth. It is a renewable resource, but we need to work together to conserve our regular use of freshwater.

Video

Aquatic Biologist

Some scientists work to conserve and protect freshwater sources such as lakes, rivers, and wetlands. These scientists are called aquatic biologists. *Aquatic* means "water." Aquatic biologists help conserve freshwater areas by protecting them from being developed by housing and business industries. This, in turn, protects freshwater areas from pollution. Aquatic biologists also do research to study freshwater areas and learn more about the organisms that live in the area.

SEP Engaging in Arguments from Evidence

Talk Together

Carlos de la Rosa is an aquatic biologist. He focuses on insects. He collects and studies insects in a wetland area of Florida. In the video segment, de la Rosa said that he can use the insects to measure the quality of the water where they live. What do you think he meant by this?

What do you think the number and variety of insects has to do with water quality in a freshwater area such as wetlands?

Pollution

Insects can be used to learn about the amount of pollution in a river or stream. The following table classifies insects based on the type of water in which they can be found.

Water Quality	Type of Insects
Almost no pollution	Caddisfly larva, mayfly nymph, dobsonfly larva
Small amounts of pollution	Damselfly nymph, dragonfly nymph, sow bugs
Medium amount of pollution	Midge larva, flatworms, blackfly larva
Lots of pollution	Blood midge larva, rat-tailed maggot

A scientist obtained water samples from two different creeks. She found the following insects in the samples:

Sample from Creek 1	Sample from Creek 2
2 midge larvae	1 damselfly nymph
2 flatworms	1 flatworm
1 red-tailed maggot	2 sow bugs
3 blackfly larvae	3 dragonfly nymphs

Analyze the data. Identify the creek with cleaner water. Use evidence from the data to support your response.

Activity 18

Evaluate Like a Scientist

Quick Code: ca5554s

Review: Water as a Valuable Natural Resource

Think about what you have read and seen in this lesson. **Write** some core ideas you have learned. Review your notes with a partner. Your teacher may also have you take a practice test.

Talk Together

Think about what you saw in Get Started. Use your new ideas to discuss water and why it is such a valuable natural resource.

SEP **Obtaining, Evaluating, and Communicating Information**

CONCEPT

3.3

Protecting Earth's Resources

Student Objectives

By the end of this lesson:

- ☐ I can describe how human activities have major effects on land, vegetation, and water reservoirs.
- ☐ I can calculate the quantity of trees required to meet a school system's paper demands.
- ☐ I can generate and compare multiple solutions to conserve Earth's natural resources.
- ☐ I can define a problem relating to overconsumption of Earth's resources.
- ☐ I can describe how humans depend on oceans and how human activities affect oceans.
- ☐ I can argue from evidence for the conservation of ocean resources.
- ☐ I can explain ways individual communities use science ideas to protect Earth's resources and environment.

Key Vocabulary

☐ biodiversity	☐ oxygen
☐ carbon dioxide	☐ plankton
☐ conserve	☐ pollute
☐ food chain	☐ recycle
☐ nonrenewable	☐ renewable resource
☐ nonrenewable resource	☐ sustainable

Quick Code:
ca5556s

Activity 1

Can You Explain?

How does conserving natural resources protect the environment?

Quick Code:
ca5558s

Activity 2

Ask Questions Like a Scientist

Quick Code: ca5559s

Disposing of Trash

Think of containers that food comes in, such as milk cartons or tubs of cottage cheese. In the table, **list** objects that you dispose of after you use them, such as plastic containers, paper, and spoiled food. Next to each item, **write** what you do with it when you finish using it. Then, **answer** the question.

Item	What I Do with It

Where do all the materials that we throw away go?

SEP **Asking Questions and Defining Problems**

Watch the video. Then, **brainstorm** different ways the land or water is affected by disposing of trash by **completing** the sentences.

Let's Investigate Disposing of Trash

I wonder ______________________________.

I wonder ______________________________.

I wonder ______________________________.

I wonder ______________________________.

Activity 3
Evaluate Like a Scientist

Quick Code: ca5560s

What Do You Already Know About Protecting Earth's Resources?

Recycle or Not?

Circle the item that should NOT be put in a recycling bin.

Natural Resources

Circle the natural resources.

- aluminum
- plastic
- sand
- water
- oil

True or False?

Circle the resources that are found in or under the ocean.

- seafood
- minerals
- freshwater
- natural gas
- petroleum

Identify Resources

Look at the images. Under each image, **write** if the resource in each image is renewable or nonrenewable and why.

Coal Mine

Oil Rig

Palm Trees

Why Is It Important That We Protect Our Natural Resources?

Activity 4

Observe Like a Scientist

Quick Code: ca5561s

Earth's Resources

Complete the first part of the activity: Gifts from Nature. Then, **complete** the table about the things you saw in the activity.

Earth's Resources

Natural Resources	Made from Natural Resources

Activity 5

Analyze Like a Scientist

Using Natural Resources Wisely

Quick Code:
ca5562s

Read the text. As you read, **highlight** evidence from the text that will help you answer the Can You Explain? question.

Using Natural Resources Wisely

It is important to use natural resources wisely. Resources that do not replace themselves in a reasonably short time are called **nonrenewable resources**. These include resources that are mined from Earth, such as minerals. Once these minerals are removed, they cannot be replaced. For example, metals are made from minerals. Some metals that are used in making electronics are very rare. Industries are already reporting shortages of these metals, including silicon.

Natural Resources

Another example of a nonrenewable resource is fossil fuels. Most of the energy we use comes from fossil fuels, such as natural gas, oil, and coal. These natural resources are also **nonrenewable**, so there is a limited amount for us to use. We will need to reduce our use of these resources and begin using other natural resources to meet the energy needs of the future. By reducing our use of fossil fuels, we can also reduce the amount of pollution that goes into our environment.

Using Natural Resources Wisely *cont'd*

Other natural resources, such as trees, food crops, fish, or water, can replace themselves over relatively short time periods. These resources are called **renewable resources**. However, as humans use more of these renewable resources, they use more than can be replaced. When this happens, shortages occur, and the price of these resources goes up. For example, if forests are cut down faster than they can grow, eventually wood will be in short supply. In some parts of the world, this has already occurred. If we eat more fish than are replaced in the ocean, fish become rarer, and fishing crews catch less. This is a problem in most of the world's oceans. In some parts of the world, people use water from wells faster than it is replaced by rain. Eventually, the water will run out, and the wells will become dry. One way to prevent this from happening is to use these resources more carefully. This is called resource conservation.

Activity 6

Observe Like a Scientist

Quick Code: ca5563s

Renewables and Nonrenewables

Watch the video and **look** for descriptions of renewable and nonrenewable resources.

Renewables and Nonrenewables

Talk Together

Now, talk together about what you saw in the video about resources.

Activity 7

Evaluate Like a Scientist

Quick Code: ca5564s

Paper Recycling

Mary is a student in elementary school. She was concerned about the amount of paper that schools are using. Mary decided she wanted to make sure that the trees used to make this paper were replaced. To do this, she needed to discover how many acres of forest were used. Her research shows the following:

- There are approximately 900 trees per acre in a forest.
- Each tree can produce two cartons of paper.
- Her school uses 18 cartons of paper each year.
- There are 2,500 schools in her state.

Calculate how many trees would need to be planted each year to replace those used to make paper. Then, **answer** the questions.

How many acres of trees would need to be planted each year? Show your calculations and explain your reasoning.

What else could Mary suggest that the school do to reduce the number of trees it uses? Use the evidence listed and at least one calculation to support your claim.

SEP Using Mathematics and Computational Thinking

Activity 8
Analyze Like a Scientist

Quick Code: ca5565s

Resource Conservation

Read the text. As you read, **complete** the chart to show examples of each type of resource conservation.

Resource Conservation

Many things we use every day are made of natural resources, both renewable and nonrenewable. Paper is made from trees. Most plastic is made from oil products. It is important to **conserve** resources, so we can make sure there will be enough when we need them. *Conserve* means to protect or preserve. How can we conserve our natural resources?

Picking Up Bottles

Did you know that renewable resources can be used up if people do not use them responsibly? Water is one example. Pollution has made a lot of Earth's water undrinkable. Pollution from burning nonrenewable resources such as coal or oil can poison the soil and cause plants and animals to die. Renewable resources can be destroyed in other ways, as well. Forests can be destroyed by deforestation when too many trees are cut down. Blowing wind and flowing water can carry away soil through erosion. We must conserve and take care of all of Earth's resources.

Resource Conservation *cont'd*

To conserve resources, remember the "three Rs." We can reduce how much we use. For example, turn off water when you are brushing your teeth. We can reuse resources and materials. Try reusing a paper cup several times rather than throwing it away after one use. We can also **recycle**, turning old materials into new ones. Americans commonly recycle paper, glass, metals, and many kinds of plastic. Do you recycle at your school?

Reduce	Reuse	Recycle

Activity 9

Observe Like a Scientist

Quick Code:
ca5566s

Natural Resources and Sustainability

Watch the video. As you watch, **complete** the graphic organizer to capture evidence of environmental problems and solutions.

Natural Resources and Sustainability

Evidence of Environmental Problems	Solutions

Activity 10

Analyze Like a Scientist

Getting to Know: Ocean Resources

Quick Code:
ca5567s

Read the text. As you read, **record** ideas for protecting the ocean from pollution.

Getting to Know: Ocean Resources

Humans need oceans, but how do we treat the oceans? Most people live their entire lives on land and may never see an ocean up close. Still, everything we do affects the planet's water. Much of the garbage produced on land ends up in oceans. So, too, do human and animal wastes and toxic chemicals from factories and other sources. These things are carried by sewers and freshwater channels to oceans. Pollution from cars and machines gets into the air and ground. Over time, this pollution works its way back into the water and to the oceans. Oil and natural gas drilling, underwater mining, and shipping in the oceans also **pollutes** the water. As large as the oceans are, they suffer the consequences of human actions.

Pollution endangers plant and animal life. Many organisms die from consuming trash and chemicals. Others suffer when their habitats are destroyed by oil spills, chemical and waste runoff, or machines (such as ships and drills). Plants and animals in ocean ecosystems are all connected. When one dies, it disrupts the **food chain**. The loss of **plankton** hurts small fish that eat them, which hurts bigger predators such as sharks. Similarly, the destruction of coral reefs harms the many organisms that live there. Many people in coastal areas depend on coral reefs for food and for other resources.

Why Is It Important to Protect Ocean Resources?

Activity 11

Evaluate Like a Scientist

Quick Code: ca5568s

Running Out of Renewables

In this photograph, trees have been cut down for humans to use. The descriptions list possible problems caused by this action. **Draw** a line from each problem to the part of the environment that it will affect most.

Water becomes polluted and undrinkable.

Animals will leave when their homes are destroyed.

Wind will blow away loose soil.

Less **oxygen** and **carbon dioxide** are recycled.

CCC Cause and Effect

Activity 12

Evaluate Like a Scientist

Quick Code: ca5569s

Conserve Ocean Resources

You have been given the assignment of writing an argument in favor of conservation of ocean resources. **Underline** all the sentences you can use as evidence to support your argument.

- Human and animal wastes often find their way through other waterways into oceans.
- More than 70 percent of Earth's surface is made up of water.
- Most water found on Earth is saltwater that people cannot drink.
- Many people depend on fish, shellfish, and other ocean animals for food.
- Oceans play a crucial part in the water cycle that moves water through Earth's systems.

SEP **Engaging in Argument from Evidence**

How Can We Protect Earth's Resources?

Activity 13

Think Like a Scientist

Quick Code: ca5570s

What Can We Do?

In this investigation, you will gather and share information about what local communities are doing to minimize the effects humans have on the environment.

SEP Obtaining, Evaluating, and Communicating Information

What materials do you need? (per group)

- Paper
- Pencils
- Research materials (such as local newspapers; Internet access and computers; individual environmental activists, local public workers, or businesspeople)

What Will You Do?

1. Using the resources provided, **research** an effect that humans have on the environment.
2. Then, **prepare** and **deliver** a presentation on your findings.

Think About the Activity

Based on your research, what are the biggest sources of the environmental problem you considered? **Explain** your answer.

Based on your research, what are some potential remedies to the environmental problem you considered?

Draw a diagram of the contributors to the environmental problem you researched and the local things being done to remedy it.

What additional questions do you have about the environmental issue you researched? **Choose** one of your questions and **explain** how you could learn information to help you answer the question.

Activity 14

Solve Problems Like a Scientist

Quick Code:
ca5572s

Drinking Water

Across the world, people use water every day—for drinking, cleaning, and bathing. Why is clean drinking water important? What advantage is there to having a technological system that cleanses drinking water?

Drinking Water

To begin thinking about this question, **watch** the video segment. Then, **complete** the activity.

Video

Water Filters

Tools

What tool is used to remove impurities from water? Circle the answer.

A) Water conditioner

B) Water filter

C) Water softener

D) Water decanter

SEP Developing and Using Models

SEP Planning and Carrying Out Investigations

Water Filtration

Follow the steps to observe how visually clean a sample of dirty water becomes after being filtered. Remember to never taste dirty water, even your filtered sample. Your teacher will provide you with the following materials:

- Charcoal
- Cotton balls
- Dirt
- Plastic bottle, 8 oz, with lid
- Plastic container, 12 oz
- Sand
- Scissors
- Water

What is the problem?

____

____

What ideas do you have to solve the problem? **Draw** and **write** about it.

How will you pick one idea to test?

How will you know your idea works?

Test your design. Draw or write to show how you tested it.

Write or draw your answers to the questions in the chart. How could you improve your design?

What Worked?	What Didn't Work?

What Could Work Better?

Activity 15

Record Evidence Like a Scientist

Quick Code: ca5571s

Disposing of Trash

Now that you have learned about protecting Earth's resources, **look** again at the video about disposing of trash. You first saw this in Wonder.

Let's Investigate Disposing of Trash

Talk Together

How can you describe disposing of trash now?

How is your explanation different from before?

SEP Constructing Explanations and Designing Solutions

Look at the Can You Explain? question. You first read this question at the beginning of the lesson.

Can You Explain?

How does conserving natural resources protect the environment?

Now, you will use your new ideas about disposing of trash to answer a question.

1. **Choose** a question. You can use the Can You Explain? question or one of your own. You can also use one of the questions that you wrote at the beginning of the lesson.

My Question

2. For help answering the above question, **use** the checklist that follows. Then, **use** the checklist to plan your scientific explanation.

- ☐ Did you include examples of conservation of natural resources?
- ☐ Does your evidence include data? If you included data, did you:
- ☐ Show the measurements?
  - ☐ Describe how you collected the data?
  - ☐ Does each piece of your evidence support your claim?

Now, **write** your scientific explanation.

Conserving natural resources protects the planet because...

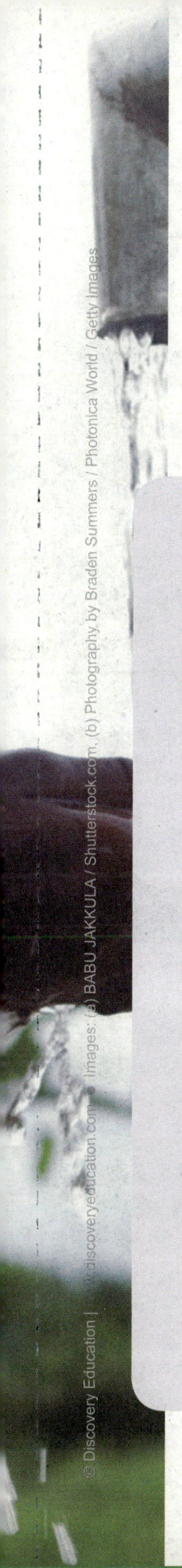

STEM in Action

Quick Code:
ca5576s

Activity 16
Analyze Like a Scientist

Wastewater Engineers

Wastewater Engineers

Using energy from the sun, Earth recycles and reuses water. But do you know that humans also move large amounts of water on Earth? Humans use, recycle, and reuse water as it moves through the water cycle. We use water to wash dishes, clean cars, and bathe dogs. We use water to brush our teeth, cook food, and even to have fun in water parks! Wherever humans are, they use water. But how do we reuse the water we have already used?

Wastewater is water that has already been used in a home, in a business, or in an industrial process. Wastewater engineers are scientists who design and make chemicals, tools, and facilities that provide us with clean water. They also check to make sure our drinking water is safe.

Wastewater Treatment

SEP Developing and Using Models

SEP Planning and Carrying Out Investigations

Wastewater Engineers *cont'd*

Some wastewater engineers work in water treatment plants. They plan the process that removes harmful materials from the water. They help choose where to build water treatment plants. These engineers observe and check each step in the process to make sure water is treated properly. They also test the treated water before people use it, and before the water is released into rivers and lakes. In this way, scientists keep the water supply safe for people and nature.

Wastewater engineers safeguard the water supply in other ways. They help to design ways to protect a community from floods. Engineers will test a community's supply of drinking water every day to check that it is safe. They may even help to design water parks so that people can enjoy the park without using too much water.

What other careers do people work in to help water move through the water cycle?

Step by Step

Based on what you have just learned, label the steps of the wastewater treatment process in order from 1 to 6.

Filters remove stones, paper, and plastic. ____________

Giant pumps force water to the surface. ____________

Grit such as sand settles at the bottom of tanks. ____________

Phosphorus is removed and sludge is formed. ____________

Water is released to rivers. ____________

Water is sprayed with coagulants. ____________

Designing a Solution

Suppose a treatment plant is required to release water that contains no more than one-half milligram per liter of phosphorus. The test results show that the water contains 3 milligrams of phosphorus per liter of water.

The wastewater treatment engineer needs to release the water into the rivers. What steps should the engineer take to solve this problem?

Analyzing Chart

Phosphorus is a pollutant that is sometimes found in wastewater. Why? Phosphorus is used in fertilizers, which many farmers use to add nutrients to the soil. Phosphorus is also in detergents and animal food.

The chart indicates other uses of phosphorus. This includes the use of phosphorus on objects made of metal, such as gates and locks, because it helps resist rusting. This also includes adding phosphorus to soft drinks to add flavor. The chart shows that the "Other Uses" of phosphorus amount to about 2 million tons of phosphorus per year. Using this information, determine how much phosphorus is used per year to make animal feed.

Phosphorus Table

Uses of Phosphorus Worldwide

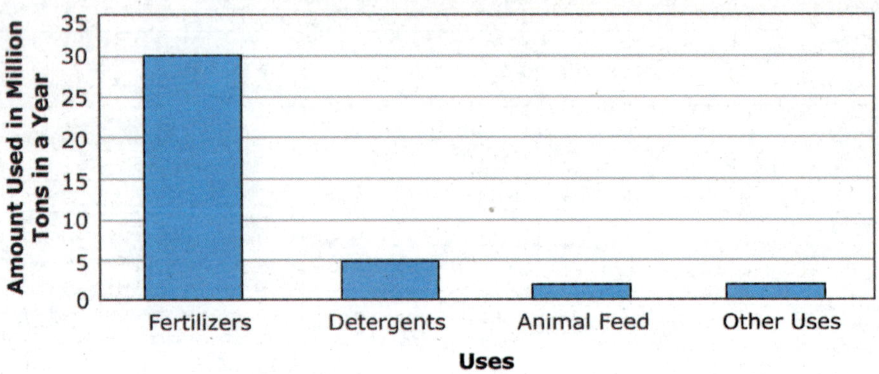

Determining Total Amount

How much phosphorus is used worldwide? How do you know?

Activity 17

Evaluate Like a Scientist

Quick Code: ca5573s

Review: Protecting Earth's Resources

Think about what you have read and seen in this lesson. **Write** some core ideas you have learned. Review your notes with a partner. Your teacher may also have you take a practice test.

SEP Obtaining, Evaluating, and Communicating Information

Talk Together

Think about what you saw in Get Started. Use your new ideas to discuss Earth's resources and how and why they should be protected.

Solve Problems Like a Scientist

Quick Code: ca5575s

Case Study: Tracking Water Pollution

In this project, you will apply your knowledge of water ecosystems as well as your math skills to analyze data about the health of seven watershed areas in Michigan.

Tracking Water Pollution

SEP Developing and Using Models

SEP Constructing Explanations and Designing Solutions

CCC Cause and Effect

Pollution is a major problem in many water ecosystems. Scientists work together to test for water pollution. Then, they design solutions for reducing it. In Michigan, a group of scientists studied seven watershed areas. The watersheds included streams and creeks. They studied the overall health of the water ecosystems by investigating the variety of water bugs in the ecosystems. The data they collected over a two-year period are shown in the graph.

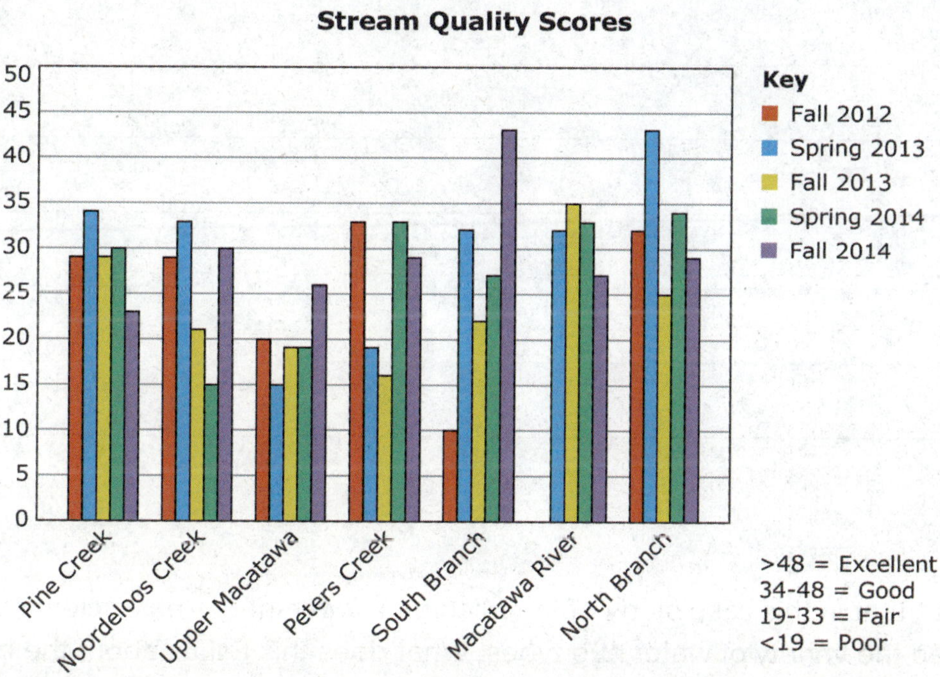

Health of Ecosystems

Look at the data. **Write** a claim about the health of the water ecosystems between Fall 2012 and Fall 2014. **Use** the data to support your claim.

Bugs

You learned that in this case study of the Michigan watershed areas, scientists investigated the variety of water bug types. What does this tell us about the health of the water ecosystems? In other words, how and why did they use water bugs to determine the health of the ecosystems? **Use** scientific reasoning to support your answer.

Water Filtration

The human population is constantly growing. There is also an increase in the number of industries that use and possibly pollute water. So, it is especially important for engineers to design useful water filtration technologies to help keep water as clean as possible. (1) Think about ways in which you might design a model of a water filtration device using common household materials. (2) What would you use as a tank to hold the filtrated water? (3) What materials would you use to filter large pieces of pollution, such as pet hair and clumps of dirt? (4) What would you use to filter the pollution that cannot be seen, such as battery acid? (5) Work with a partner to **brainstorm** ideas. Then, **make** a model of your filtration device in the form of a labeled sketch. Be sure to **include** information about the materials you would like to use in the model and how they would work.

Think About the Activity

After you have completed your model, **discuss** with a partner the benefits and limitations of your model and how you could improve it. **Write** or **draw** your ideas in the chart.

What Worked?	What Didn't Work?

What Could Work Better?

Grade 5 Resources

Name ______________________________

Bubble Map

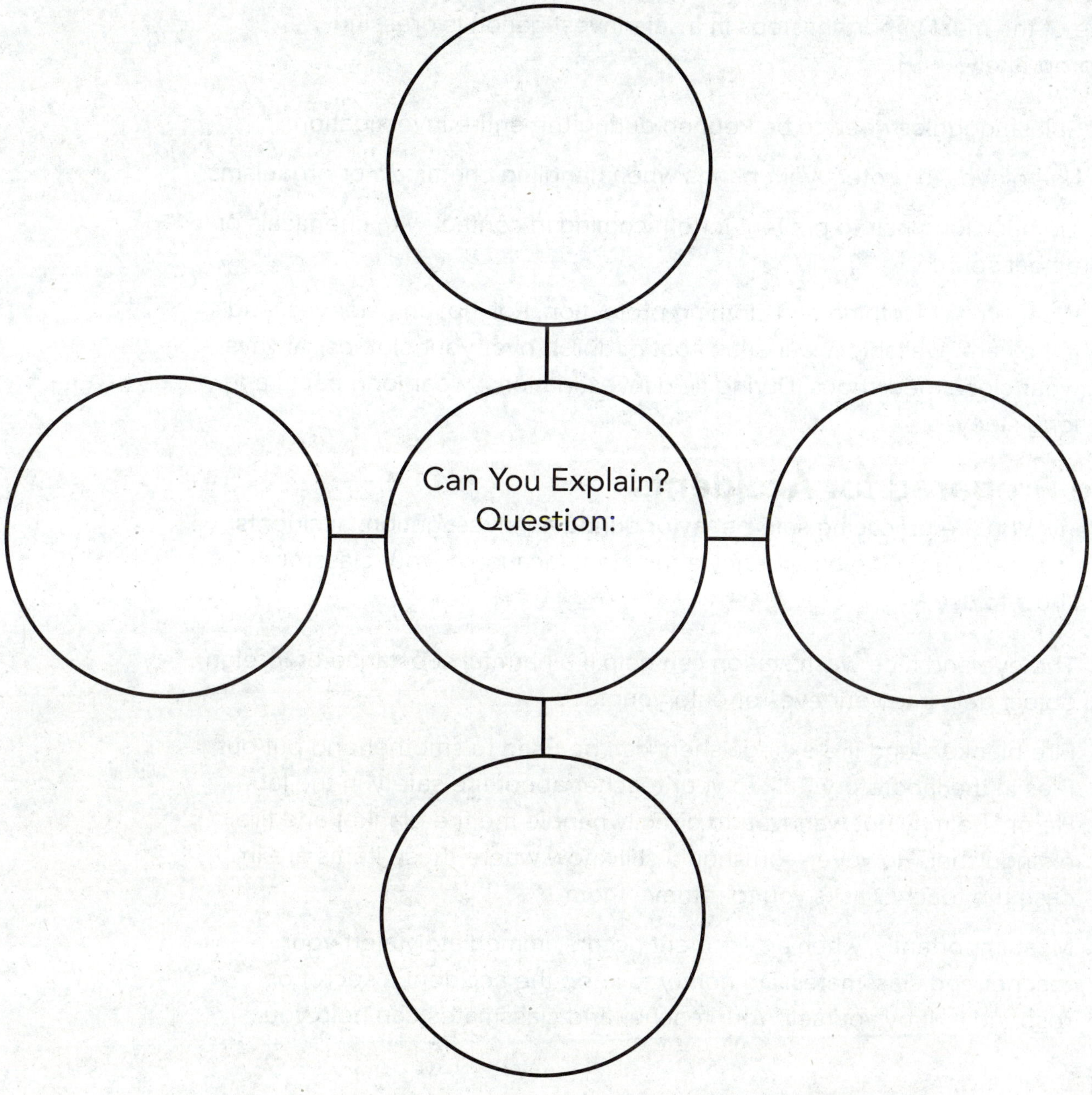

Safety in the Science Classroom

Following common safety practices is the first rule of any laboratory or field scientific investigation.

Dress for Safety

One of the most important steps in a safe investigation is dressing appropriately.

- Splash goggles need to be kept on during the entire investigation.
- Use gloves to protect your hands when handling chemicals or organisms.
- Tie back long hair to prevent it from coming in contact with chemicals or a heat source.
- Wear proper clothing and clothing protection. Roll up long sleeves, and if they are available, wear a lab coat or apron over your clothes. Always wear close toed shoes. During field investigations, wear long pants and long sleeves.

Be Prepared for Accidents

Even if you are practicing safe behavior during an investigation, accidents can happen. Learn the emergency equipment location in your classroom and how to use it.

- The eye and face wash station can help if a harmful substance or foreign object gets into your eyes or onto your face.
- Fire blankets and fire extinguishers can be used to smother and put out fires in the laboratory. Talk to your teacher about fire safety in the lab. He or she may not want you to directly handle the fire blanket and fire extinguisher. However, you should still know where these items are in case the teacher asks you to retrieve them.
- Most importantly, when an accident occurs, immediately alert your teacher and classmates. Do not try to keep the accident a secret or respond to it by yourself. Your teacher and classmates can help you.

Practice Safe Behavior

There are many ways to stay safe during a scientific investigation. You should always use safe and appropriate behavior before, during, and after your investigation.

Safety Goggles

- Read the all of the steps of the procedure before beginning your investigation. Make sure you understand all the steps. Ask your teacher for help if you do not understand any part of the procedure.
- Gather all your materials and keep your workstation neat and organized. Label any chemicals you are using.
- During the investigation, be sure to follow the steps of the procedure exactly. Use only directions and materials that have been approved by your teacher.
- Eating and drinking are not allowed during an investigation. If asked to observe the odor of a substance, do so using the correct procedure known as wafting, in which you cup your hand over the container holding the substance and gently wave enough air toward your face to make sense of the smell.
- When performing investigations, stay focused on the steps of the procedure and your behavior during the investigation. During investigations, there are many materials and equipment that can cause injuries.
- Treat animals and plants with respect during an investigation.
- After the investigation is over, appropriately dispose of any chemicals or other materials that you have used. Ask your teacher if you are unsure of how to dispose of anything.
- Make sure that you have returned any extra materials and pieces of equipment to the correct storage space.
- Leave your workstation clean and neat. Wash your hands thoroughly.

abiotic

Image: Paul Fuqua

includes all nonliving things

abyssal zone

Image: Public Domain Pictures/Pixabay

the zone of deep ocean below a depth of 4,000 meters

aquatic

Image: Paul Fuqua

relating to water

Arctic

Image: Discovery Communications, Inc.

being from an icy climate, such as the North Pole

atmosphere

Image: Paul Fuqua

layers of gas that surround a planet

biodiversity

Image: Discovery Communications, Inc.

all of the organisms that live together in an environment

biosphere

Image: Grisha Bruev/Shutterstock

that part of Earth in which life can exist

biotic

Image: Paul Fuqua

includes all living things

carbon dioxide

Image: Steve Buissinne/Pixabay

a waste product made by cells of the body; a gas in the air made of carbon and oxygen atoms

conserve

Image: Emilian Robert Vicol/Pixabay

to protect something, or prevent the wasteful overuse of a resource

crater

Image: NASAa

a large, circular pit in the surface of a planet or other body in space usually formed when two bodies in space collide

Earth

Image: NASAa

the third planet from the Sun; the planet on which we live

ecosystem

Image: Paul Fuqua

all the living and nonliving things in an area that interact with each other

estuary

Image: Discovery Communications, Inc.

a coastal body of water where fresh water from a river mixes with salt water from the ocean

food chain

Image: Ocean First Education

a model that shows one set of feeding relationships among living things

freshwater

Image: Paul Fuqua

water that is not salty, such as that found in streams and lakes

geosphere

Image: Thorin Wolfheart / Shutterstock.com

Earth's crust, both beneath the oceans and continents, as well as the mantle and inner and outer core

glacier

Image: Discovery Communications, Inc.

a large sheet of ice or snow that moves slowly over Earth's surface

hydrosphere

Image: Bernhard Starck/Pixabay

all of the water on, under, and above the Earth

mineral

Image: Paul Fuqua

a natural, solid substance found in rocks; each mineral has a specific chemical makeup

natural resources

Image: Paul Fuqua

resources that are obtained from the earth

nonrenewable

Image: Paul Fuqua

once it is used, it cannot be made or reused again

nonrenewable resource

Image: Paul Fuqua

a natural resource of which a finite amount exists, or one which cannot be replaced with currently available technologies

ocean

Image: Martin Winkler/Pixabay

a large body of salt water that covers most of Earth

organism

Image: Discovery Communications, Inc.

any individual living thing

oxygen

Image: Paul Fuqua

a gas in Earth's atmosphere and in water that living organisms breathe

phenomena

Image: David Mark/Pixabay

a scientific wonder or happening

plankton

Image: Paul Fuqua

small organisms that drift through bodies of water; include animals, plants, and bacteria

pollute

Image: Paul Fuqua

to put harmful materials into the air, water, or soil (related words: pollution, pollutant)

precipitation

Image: Paul Fuqua

water that is released from clouds in the sky; includes rain, snow, sleet, hail, and freezing rain

recycle

Image: Rawpixel.com/Shutterstock

to create new materials from used products

refuse (n)

Image: Rawpixel.com/Shutterstock

garbage

relies

Image: skeeze/Pixabay

dependent upon

renewable resource

Image: Paul Fuqua

a natural resource that can be replaced

resource

Image: Pexels/Pixabay

a naturally occurring material in or on Earth's crust or atmosphere of potential use to humans

river

Image: Paul Fuqua

big body of water flowing through land on either side

salt

Image: andreas160578/Pixabay

a mineral found in the ocean and other parts of Earth that can be used for preserving things and seasoning food

saltwater

Image: Paul Fuqua

a general term for water containing more than 1 part per thousand (1 g/L) of total dissolved solids

stream

Image: Paul Fuqua

a small body of flowing water

substance

Image: Olga Chesnokova/Shutterstock

the physical matter of which living or nonliving things are composed

sustainable

Image: PublicDomainPictures/Pixabay

able to be used over and over again without hurting the overall supply

terrarium

Image: Pixabay

a man-made ecosystem used for organisms to grow, usually in a clear container to allow for observation

volcano

Image: MonikaP/Pixabay

an opening in the Earth's surface through which magma and gases or only gases erupt

water

Image: Paul Fuqua

a compound made of hydrogen and oxygen

watershed

Image: Brian Cluer NOAA Fisheries West Coast Region California Coastal Office

a region in which all precipitation and surface water collects and drains into the same river

English	A	Español
abiotic includes all nonliving things		**abiótico** incluye todos los objetos sin vida
absorb to take in		**absorber** incorporar
abyssal zone the zone of deep ocean below a depth of 4,000 meters		**zona abisal** zona del océano profundo por debajo de los 4,000 metros de profundidad
air the part of the atmosphere closest to Earth; the part of the atmosphere that organisms on Earth use for respiration		**aire** parte de la atmósfera más cercana a la Tierra; la parte de la atmósfera que los organismos que habitan la Tierra utilizan para respirar
aquatic relating to water		**acuático** relativo al agua
Arctic being from an icy climate, such as the North Pole		**ártico** que pertenece a un clima helado, como el Polo Norte

astronaut
a person who travels outside Earth's atmosphere; a person who travels into space (related words: astronomer, astronomy)

astronauta
persona que viaja fuera de la atmósfera de la Tierra; una persona que viaja por el espacio (palabras relacionadas: astrónomo, astronomía)

---

atmosphere
layers of gas that surround a planet (related word: atmospheric)

atmósfera
capas de gas que rodean un planeta (palabra relacionada: atmosférico)

---

axis
an imaginary line that an object spins or revolves around

eje
línea real o imaginaria que pasa por el centro de un objeto; el objeto gira alrededor de ella

B

beaker
a scientific measuring cup used to measure liquids

vaso de precipitados
contenedor cilíndrico de vidrio usado en laboratorio

biodiversity
all of the organisms that live together in an environment

biodiversidad
todos los organismos que viven juntos en un medio ambiente

---

biosphere
that part of Earth in which life can exist

biosfera
parte de la Tierra donde puede existir la vida

---

biotic
includes all living things

biótico
incluye a todos los seres vivos

---

boil
to change state from a liquid to a gas because of added heat

hervir
cambiar el estado de líquido a gas por el aumento de la temperatura

C

carbon dioxide
a waste product made by cells of the body; a gas in the air made of carbon and oxygen atoms: humans rid themselves of carbon dioxide waste by exhaling, or breathing out

dióxido de carbono
desecho formado por células del cuerpo; gas en el aire que se forma por átomos de carbono y oxígeno: los humanos desechamos dióxido de carbono al exhalar o expulsar aire

celestial sphere an imaginary sphere surrounding Earth that show the sky and stars overhead	**esfera celeste** esfera imaginaria que rodea la Tierra, y que muestra el cielo y las estrellas
Celsius the metric temperature scale: Water boils at 100 degrees Celsius, and it freezes at 0 degrees Celsius	**Celsius** escala de la temperatura métrica: el agua hierve a los 100 grados centígrados y se congela a los 0 grados centígrados
change of state the process of changing (by heating or cooling) one form of matter into another form of matter with different characteristics	**cambio de estado** proceso de cambiar (mediante frío o calor) una forma de materia a otra forma de materia con diferentes características
chemical change a process that changes substances into new substances and is different physically from the original substance	**cambio químico** proceso que transforma a las sustancias en nuevas sustancias que son diferentes físicamente de la que las originó (palabra relacionada: reacción química)

chemical energy
energy that can be changed into motion and heat, stored in the bonds between atoms

energía química
energía que se puede cambiar a movimiento y calor, y que está almacenada en las cadenas entre átomos

---

combine
to bring together; to mix (related word: combination)

combinar
unir, mezclar (palabra relacionada: combinación)

---

conserve
to protect something or prevent the wasteful overuse of a resource

conservar
proteger algo o evitar el uso excesivo e ineficiente de un recurso

---

constant
continuing without interruption

constante
que continúa sin interrupción

---

constellation
a particular area of the sky; a group of stars

constelación
área particular del cielo; grupo de estrellas

consumer
an organism that eats other living things to get energy; an organism that does not produce its own food (related word: consume)

consumidor
organismo que come otros seres vivos para obtener energía; organismo que no produce su propio alimento (palabra relacionada: consumir)

---

crater
a large, circular pit in the surface of a planet or other body in space usually formed when two bodies in space collide

cráter
hoyo grande y circular en la superficie de un planeta u otro cuerpo en el espacio, generalmente formado cuando chocan dos cuerpos en el espacio

---

cycle
a process that repeats (related word: cyclic)

ciclo
proceso que se repite (palabra relacionada: cíclico)

D

decomposer
organism that carries out the process of breaking things down into dead or decaying organisms

descomponedor
organismo que lleva a cabo el proceso de descomposición mediante la desintegración de los organismos muertos

E

Earth the third planet from the sun; the planet on which we live (related words: earthly; earth – meaning soil or dirt)	**Tierra** tercer planeta desde el Sol; planeta en el cual vivimos (palabras relacionadas: terrenal; tierra en el sentido de suelo o suciedad)
ecosystem all the living and nonliving things in an area that interact with one another	**ecosistema** todos los seres vivos y objetos sin vida de un área, que se interrelacionan entre sí
electron a particle with a negative charge	**electrón** partícula subatómica con una carga negativa
energy the ability to do work or cause change; the ability to move an object some distance	**energía** habilidad para hacer un trabajo o producir un cambio; habilidad para mover un objeto a cierta distancia

energy pyramid
a model that represents energy transfer within an ecosystem, showing a lot of energy at the bottom and a much smaller amount on top

pirámide energética
modelo que representa la transferencia de energía dentro de un ecosistema, que muestra mucha energía en la base y una cantidad mucho más pequeña en la cima

---

energy transfer
the transfer of energy from one organism to another through a food chain or web; or the transfer of energy from one object to another, such as heat energy

transferencia de energía
transmisión de energía desde un organismo a otro a través de una cadena o red de alimentos; o transferencia de energía desde un objeto a otro, como por ejemplo la energía del calor

---

environment
all the living and nonliving things that surround an organism

medio ambiente
todos los seres vivos y objetos sin vida que rodean a un organismo

---

estuary
a coastal body of water where freshwater from a river mixes with saltwater from the ocean

estuario
cuerpo costero de agua donde el agua dulce de un río se mezcla con el agua salada del océano

F

food chain a model that shows how energy passes from one organism to another in an ecosystem	**cadena alimentaria** modelo que muestra un conjunto de relaciones de alimentación entre seres vivos en un ecosistema
food web shows the ways in which many food chains work with one another in an ecosystem	**red alimentaria** muestra la manera en que muchas cadenas alimentarias trabajan entre sí en un ecosistema
force a pull or push that is applied to an object	**fuerza** acción de atraer o empujar que se aplica a un objeto
freshwater water that is not salty, such as that found in streams and lakes	**agua dulce** agua que no es salada, como la que se encuentra en arroyos y lagos

friction
a force that stops motion

fricción
fuerza que desacelera o detiene el movimiento

G

galaxy
a group of solar systems, dust, and gas held together by gravity; our solar system is part of the Milky Way galaxy

galaxia
grupo de sistemas solares, polvo y gas unidos por la gravedad; nuestro sistema solar es parte de la galaxia llamada Vía Láctea

gas
a state of matter without any defined volume or shape (related word: gaseous)

gas
estado de la materia sin volumen ni forma definidos (palabra relacionada: gaseoso)

geosphere
Earth's crust, both beneath the oceans and continents, as well as the mantle and inner and outer core

geosfera
corteza terrestre, tanto debajo de los océanos como de los continentes, así como también el manto y los núcleos interior y exterior

glacier
a large sheet of ice or snow that moves slowly over Earth's surface

glaciar
capa de hielo o nieve que se mueve lentamente sobre la superficie de la Tierra

---

gravity
the force that pulls an object toward the center of Earth (related word: gravitational)

gravedad
fuerza que empuja a un objeto hacia el centro de la Tierra (palabra relacionada: gravitacional)

H

heat
the transfer of thermal energy

calor
transferencia de energía térmica

---

horizon
the point at which Earth's surface appears to meet the sky

horizonte
punto en el cual la superficie de la Tierra parece reunirse con el cielo

---

hydrogen
the most abundant element in the universe, made of one proton and one electron

hidrógeno
el elemento químico más abundante en el universo, hecho de un protón y un electrón

hydrosphere
all of the water on, under, and above Earth

hidrósfera
toda el agua que se encuentra sobre, debajo y en la Tierra

I

imaginary
existing only in your mind or in your imagination

imaginario
que existe sólo en nuestra mente o en nuestra imaginación

interact
to act on one another (related word: interaction)

interactuar
ejercer influencia mutua (palabra relacionada: interacción)

L

light
a form of energy that moves in waves and particles and can be seen

luz
una forma de energía que se mueve en ondas y partículas y que se puede ver

light energy
that form of energy that animals can see directly; visible electromagnetic radiation

energía lumínica
tipo de energía que los animales pueden ver directamente; radiación electromagnética visible

light year
the distance light travels in a vacuum in one year; about 6 trillion miles

año luz
distancia que viaja la luz en el espacio en un año; alrededor de 6 billones de millas

liquid
a state of matter with a defined volume but no defined shape

líquido
estado de la materia que posee un volumen definido, pero no una forma definida

location
a place where something is positioned

ubicación
lugar donde se posiciona algo

M

magnet
an object with a north and south pole that produces a magnetic field (related terms: magnetism, magnetic)

imán
objeto con un polo norte y un polo sur que produce un campo magnético (palabras relacionadas: magnetismo, magnético)

magnify to make something appear larger, usually by using one or more lenses	**ampliar** hacer que algo parezca más grande, generalmente usando una o más lentes
mass the amount of matter in an object	**masa** cantidad de materia en un objeto
material things that can be used to build or create something	**material** cosas que se pueden usar para construir o crear algo
matter material that has mass and takes up some amount of space	**materia** material que tiene masa y ocupa cierta cantidad de espacio
measure to use a tool to learn more about the volume, length, or weight of an object (related word: measurement)	**medir** usar una herramienta para saber más sobre el volumen, la longitud, o el peso de un objeto (palabra relacionada: medición)

mechanical energy energy that an object has because of its motion or its position	**energía mecánica** energía que tiene un objeto debido a su movimiento o posición
melt to change a substance from solid to liquid	**derretirse** cambio de sustancia de estado sólido a líquido
mineral a natural, solid substance found in rocks; each mineral has a specific chemical makeup	**mineral** sustancia natural y sólida que se encuentra en las rocas; cada mineral tiene una composición química específica
mixture a combination of substances that can be physically separated from one another	**mezcla** combinación de sustancias que pueden separarse físicamente unas de otras
model a drawing, object, or idea that represents a real event, object, or process	**modelo** dibujo, objeto o idea que representa un evento, objeto, o proceso real

moon
a body in outer space that orbits a planet; a natural satellite

luna
cuerpo en el espacio exterior que gira alrededor de un planeta; satélite natural

---

motion
a change in the position of an object compared to another object (related terms: move, movement)

movimiento
cambio en la posición de un objeto en comparación con otro objeto (palabras relacionadas: mover, desplazamiento)

N

natural resources
resources that are obtained from Earth

recursos naturales
recursos obtenidos de la Tierra

---

nebula
an interstellar cloud made up of hydrogen gas, plasma, helium gas, and dust

nebulosa
nube interestelar constituida por gases (hidrógeno y helio), plasma, y polvo

---

nonrenewable
once it is used, it cannot be made or reused again

no renovable
una vez que se usa no se puede utilizar de nuevo

nonrenewable resource
a natural resource of which a finite amount exists, or one that cannot be replaced with currently available technologies

recurso no renovable
recurso natural del cual existe una cantidad finita, o uno que no puede remplazarse con las tecnologías actualmente disponibles

nuclear energy
the energy released when the nucleus of an atom is split apart (during fission) or combined with another nucleus (during fusion)

energía nuclear
energía liberada cuando el núcleo de un átomo se divide (durante la fisión) o combina con otro átomo (durante la fusión)

nucleus
the center of an atom containing protons and neutrons (related terms: nuclei, nuclear); a region in a cell that is surrounded by a membrane and contains genetic material

núcleo
centro de un átomo que contiene protones y neutrones (palabras relacionadas: núcleos, nuclear); región de una célula que está rodeada por una membrana y contiene material genético

nutrients
important particles found in food that a living thing needs to survive

nutrientes
importantes partículas que se encuentran en los alimentos y que un ser vivo necesita para sobrevivir

O

ocean a large body of saltwater that covers most of Earth	**océano** gran cuerpo de agua salada que cubre la mayor parte de la Tierra
optical having to do with the eye or lenses	**óptico** relacionado con los ojos o con las lentes
orbit the circular path of an object as it revolves around another object	**órbita** trayectoria circular de un objeto que se forma a medida que gira alrededor de otro objeto
organism any individual living thing	**organismo** todo ser vivo individual
oxygen a gas found in air and water that living things need to breathe	**oxígeno** gas que se encuentra en el aire y en el agua, y que los organismos vivos respiran

P

particle something that is very tiny	**partícula** algo que es muy pequeño
perpendicular a downward direction, making a right angle	**perpendicular** con dirección descendente, formando un ángulo recto
phenomena a scientific wonder or happening	**fenómeno** una maravilla o suceso científico
physical change a change in matter that does not change the makeup of a substance	**cambio físico** alteración en la materia que no afecta su composición química
plankton small organisms that drift through bodies of water; include animals, plants, and bacteria	**plancton** organismos pequeños que se mueven a través de cuerpos de agua; incluyen animales, plantas, y bacterias

plant
an organism that is made up of many cells, makes its own food through photosynthesis, and cannot move; a member of kingdom Plantae

planta
organismo formado por muchas células que fabrica su propio alimento a través de la fotosíntesis, y no se puede mover; miembro del reino vegetal

---

plasma
a fourth state of matter in which the particles of a gas become highly charged (ionized)

plasma
cuarto estado de la materia en el que las partículas de un gas llegan a estar altamente cargadas (ionizadas)

---

pollute
to put harmful materials into the air, water, or soil (related words: pollution, pollutant)

contaminar
poner materiales perjudiciales en el aire, agua, o suelo (palabras relacionadas: contaminación, contaminante)

---

precipitation
water that is released from clouds in the sky; includes rain, snow, sleet, hail, and freezing rain

precipitación
agua liberada de las nubes en el cielo; incluye la lluvia, la nieve, la aguanieve, el granizo, y la lluvia congelada

prey
an animal that is hunted and eaten by another animal

presa
animal que es cazado y comido por otro

---

produce (v)
to make or create something

producir (v)
hacer o crear algo

---

producer
an organism that makes its own food; an organism that does not consume other plants or animals

productor
organismo que fabrica su propio alimento; organismo que no consume otras plantas u otros animales

---

property
a characteristic or quality of a material

propiedad
característica o calidad de un material

R

radiant energy
energy that does not need matter to travel; light

energía radiante
energía que no necesita de la materia para viajar; luz

---

radiation
electromagnetic energy (related word: radiate)

radiación
[update for this definition tk] (palabra relacionada: irradiar)

English	Spanish
recycle to create new materials from used products	**reciclar** crear nuevos materiales a partir de productos usados
refuse (n) garbage	**desecho (s)** basura
relies dependent upon	**depender** [update for this definition tk]
renewable resource a natural resource that can be replaced	**recurso renovable** recurso natural que puede reemplazarse
resource a naturally occurring material in or on Earth's crust or atmosphere of potential use to humans	**recurso** material que se origina de forma natural en o sobre la corteza o la atmósfera de la Tierra, que es de uso potencial para los seres humanos
revolution the orbiting of an object around another object	**revolución** movimiento por el cual un objeto gira alrededor de otro objeto describiendo una órbita completa

river
big body of water flowing through land on either side

río
gran cuerpo de agua que fluye con tierra de ambos lados

---

rotate
turning around on an axis; spinning (related word: rotation)

rotar
girar sobre un eje; dar vueltas (palabra relacionada: rotación)

---

rotation
the spinning of a celestial body, such as a planet, around an axis

rotación
giro de un cuerpo celeste, como un planeta, alrededor de un eje

S

salt
a mineral found in the ocean and other parts of Earth that can be used for preserving things and seasoning food

sal
un mineral que se encuentra en el océano y otras partes de la Tierra, que se puede usar para conservar cosas y condimentar alimentos

---

saltwater
contains salt and other minerals that make the water unsuitable for drinking

agua salada
contiene sal y otros minerales que hacen que el agua no sea apta para beber

scale
a device used for measuring weight

escala/báscula
dispositivo empleado para medir el peso

---

shelter
a location or structure that gives protection from danger or bad weather

refugio
lugar o estructura que brinda protección contra los peligros o las malas condiciones meteorológicas

---

solar system
a system of objects that revolve around a star

sistema solar
conjunto de objetos que giran alrededor de una estrella

---

solid
matter with a fixed volume and shape

sólido
materia con un volumen y una forma determinada

sound
anything you can hear that travels by making vibrations in air, water, and solids

sonido
vibración que viaja a través de un material, como el aire o el agua

---

star
a massive ball of gas in outer space that gives off heat, light, and other forms of radiation

estrella
bola masiva de gas en el espacio exterior que emite calor, luz y otras formas de radiación

---

state of matter
a particular form that matter can take: the three main states of matter are solid, liquid, and gas.

estado de la materia
forma particular que puede tener la materia: los tres estados principales de la materia son sólido, líquido, y gaseoso

---

stem
the part of a plant that grows away from the roots; supports leaves and flowers

tallo
parte de la planta que crece a partir de la raíz; contiene las hojas y las flores

stomata pores on the surface of a plant that allow gases to move into and out of the plant (related word: stoma)	**estomas** poros en la superficie de una planta que permiten a los gases moverse hacia adentro y hacia afuera de la planta (palabra relacionada: estoma)
stream a small body of flowing water	**arroyo** pequeño cuerpo de agua que fluye
substance the physical matter of which living or nonliving things are made	**sustancia** materia física de la cual están compuestos los seres vivos y objetos sin vida
sugar a substance that is sweet, mainly used in food and drinks	**azúcar** sustancia dulce, utilizada principalmente en alimentos y bebidas
sun any star around which planets revolve	**sol** toda estrella alrededor de la cual giran los planetas

surface the top or outermost part of something	**superficie** parte superior o externa de un objeto
survive to continue living or existing: an organism survives until it dies; a species survives until it becomes extinct (related word: survival)	**sobrevivir** continuar viviendo o existiendo: un organismo sobrevive hasta que muere; una especie sobrevive hasta que se extingue (palabra relacionada: supervivencia)
sustainable able to be used over and over again without hurting the overall supply	**sostenible** que se puede utilizar una y otra vez sin afectar el suministro total
system a group of related objects that work together to perform a function	**sistema** grupo de objetos relacionados que funcionan juntos para realizar una función

T

telescope
an instrument used to observe objects that are far away

telescopio
instrumento usado para observar objetos que se encuentran alejados

terrarium
a human-made ecosystem used for organisms to grow, usually in a clear container to allow for observation

terrario
ecosistema creado por el hombre para que crezcan organismos, generalmente en un recipiente transparente para permitir la observación

thermal energy
energy in the form of heat

energía térmica
energía en forma de calor

U

universe
everything that exists in, on, and around Earth

universo
todo lo que existe en, sobre, o alrededor de la Tierra

V

volcano
an opening in Earth's surface through which magma and gases or only gases erupt (related word: volcanic)

volcán
abertura en la superficie de la Tierra a través de la cual el magma y los gases o sólo los gases hacen erupción (palabra relacionada: volcánico)

---

volume
the amount of space that matter takes up

volumen
la cantidad de espacio que ocupa la materia

W

warm
having heat

cálido
que tiene calor

---

water
a compound made of hydrogen and oxygen; can be in either a liquid, ice, or vapor form and has no taste or smell

agua
compuesto hecho de hidrógeno y oxígeno, puede presentarse en forma líquida, en forma de hielo o vapor y no tiene sabor ni olor

watershed
a region in which all precipitation and surface water collects and drains into the same river

cuenca
región en la que se recoge toda la precipitación y agua superficial, que drena hacia el mismo río

---

weight
the force of gravity on an object

peso
fuerza de gravedad que se ejerce sobre un objeto

Index